SURGERY - PROCEDURES, COMPLICATIONS, AND RESULTS SERIES

STEM CELLS AND CARTILAGE TISSUE ENGINEERING APPROACHES TO ORTHOPAEDIC SURGERY

SURGERY – PROCEDURES, COMPLICATIONS, AND RESULTS SERIES

Stem Cells and Cartilage Tissue Engineering Approaches to Orthopaedic Surgery
Wasim S Khan, Timothy E Hardingham
2010. ISBN: 978-1-60876-864-6

Surgical Simulation and Training
Jamie L. Huang (Editor)
2010. ISBN: 978-1-61668-437-2

Surgical Simulation and Training
Jamie L. Huang (Editor)
2010. ISBN: 978-1-61668-465-5 (Online Book)

Hand Surgery: Preoperative Expectations, Techniques and Results
Robert H. Beckingsworth (Editor)
2010. ISBN: 978-1-60876-280-4

Regional Outcome Measure in Hand Surgery
Wasim S Khan and Matthew A Ravenscroft
2010. ISBN: 978-1-60876-685-7

Gastric Bypass: Surgical Procedures, Health Effects and Common Complications
Jennifer T. Rogers (Editor)
2010. ISBN: 978-1-60876-698-7

SURGERY - PROCEDURES, COMPLICATIONS, AND RESULTS SERIES

STEM CELLS AND CARTILAGE TISSUE ENGINEERING APPROACHES TO ORTHOPAEDIC SURGERY

WASIM S. KHAN
AND
TIMOTHY E. HARDINGHAM
EDITORS

Nova Science Publishers, Inc.
New York

Copyright © 2010 by Nova Science Publishers, Inc.

All rights reserved. No part of this book may be reproduced, stored in a retrieval system or transmitted in any form or by any means: electronic, electrostatic, magnetic, tape, mechanical photocopying, recording or otherwise without the written permission of the Publisher.

For permission to use material from this book please contact us:
Telephone 631-231-7269; Fax 631-231-8175
Web Site: http://www.novapublishers.com

NOTICE TO THE READER

The Publisher has taken reasonable care in the preparation of this book, but makes no expressed or implied warranty of any kind and assumes no responsibility for any errors or omissions. No liability is assumed for incidental or consequential damages in connection with or arising out of information contained in this book. The Publisher shall not be liable for any special, consequential, or exemplary damages resulting, in whole or in part, from the readers' use of, or reliance upon, this material.

Independent verification should be sought for any data, advice or recommendations contained in this book. In addition, no responsibility is assumed by the publisher for any injury and/or damage to persons or property arising from any methods, products, instructions, ideas or otherwise contained in this publication.

This publication is designed to provide accurate and authoritative information with regard to the subject matter covered herein. It is sold with the clear understanding that the Publisher is not engaged in rendering legal or any other professional services. If legal or any other expert assistance is required, the services of a competent person should be sought. FROM A DECLARATION OF PARTICIPANTS JOINTLY ADOPTED BY A COMMITTEE OF THE AMERICAN BAR ASSOCIATION AND A COMMITTEE OF PUBLISHERS.

LIBRARY OF CONGRESS CATALOGING-IN-PUBLICATION DATA

Stem cells and cartilage tissue engineering approaches to orthopaedic surgery / Wasim S. Khan, Timothy E. Hardingham.
   p. ; cm.
  Includes bibliographical references and index.
  ISBN 978-1-60876-864-6 (softcover)
 1. Orthopedics. 2. Stem cells--Transplantation. 3. Tissue engineering. I. Hardingham, Timothy E. II. Title.
  [DNLM: 1. Stem Cell Transplantation. 2. Stem Cells--physiology. 3. Cartilage--physiology. 4. Orthopedic Procedures. 5. Tissue Engineering. QU 325 K45s 2009]
  RD732.K53 2009
  616'.02774--dc22
                          2009050567

Published by Nova Science Publishers, Inc. † New York

CONTENTS

Preface		vii
Abstract		ix
Chapter 1	Stem Cells	1
Chapter 2	Sources of Adult Mesenchymal Stem Cells	5
Chapter 3	Differentiation of Stem Cells	9
Chapter 4	Stem Cell Markers	13
Chapter 5	Articular Cartilage	21
Chapter 6	Current Surgical Treatment Modalities	25
Chapter 7	Tissue Engineering Approaches for Articular Cartilage Defects	29
Conclusions		33
References		35
Index		49

PREFACE

Tissue is frequently damaged or lost in injury and disease. There has been an increasing interest in stem cell applications and tissue engineering approaches in surgical practice to deal with damaged or lost tissue. Tissue engineering is an exciting strategy being explored to deal with damaged or lost tissue. It is the science of generating tissue using molecular and cellular techniques, combined with material engineering principles, to replace tissue. This could be in the form of cells with or without matrices. Although there have been developments in almost all surgical disciplines, the greatest advances are being made in orthopaedics, especially in cartilage repair. This is due to many factors including the familiarity with bone marrow derived mesenchymal stem cells and cartilage being a relatively simpler tissue to engineer. Unfortunately significant hurdles remain to be overcome in many areas before tissue engineering becomes more routinely used in clinical practice.

ABSTRACT

Tissue is frequently damaged or lost in injury and disease. There has been an increasing interest in stem cell applications and tissue engineering approaches in surgical practice to deal with damaged or lost tissue. Tissue engineering is an exciting strategy being explored to deal with damaged or lost tissue. It is the science of generating tissue using molecular and cellular techniques, combined with material engineering principles, to replace tissue. This could be in the form of cells with or without matrices. Although there have been developments in almost all surgical disciplines, the greatest advances are being made in orthopaedics, especially in cartilage repair. This is due to many factors including the familiarity with bone marrow derived mesenchymal stem cells and cartilage being a relatively simpler tissue to engineer. Unfortunately significant hurdles remain to be overcome in many areas before tissue engineering becomes more routinely used in clinical practice.

Cells used in tissue engineering could be autologous, allogeneic or xenogeneic. The cells could be stem cells or cells further down the differentiation pathway. The use of embryonic stem cells is associated with religious, political and social concerns, but the use of adult stem cells is generally well accepted. Stem cells have been identified in a number of adult tissues, albeit in small numbers. In addition to bone marrow, mesenchymal stem cells have been identified in a number of tissues including adipose tissue and fat pad. The mesenchymal stem cells are generally isolated from the tissue and expanded in culture.

These cells are characterised or defined using a set of cell surface markers; mesenchymal stem cells are generally positive for CD44, CD90 and CD105, and are negative for haematopoetic markers CD34 and CD45, and the neurogenic marker CD56.

These cells can be differentiated down a particular differentiation pathway e.g. osteoblast or chondrocyte, using predefined culture conditions before being used for clinical applications. In this book stem cells are discussed in detail followed by the tissue engineering approaches for articular cartilage.

Chapter 1

STEM CELLS

Orthopaedic surgery has been highly successful in repairing, realigning and replacing damaged musculoskeletal structures. The coming years will establish whether a paradigm shift from fixation towards regeneration of tissue is possible, clinically feasible and financially viable. In developing tissue engineering techniques based on mesenchymal stem cells, a better understanding of these cells, their various sources, their characterisation and their differentiation potential will be crucial in devising personalised treatment strategies suited to individual patients and lesions. Without this knowledge, there is a risk that suboptimal cell-based treatments will fare misleadingly badly in comparative clinical studies.

Stem cells are a self-renewing, slow-cycling cell population that exhibit high clonogenity, low cellular proliferation and the ability to undergo multilineage differentiation. Humans originate from the ultimate stem cell, the fertilised egg, and develop through a process of cell proliferation and differentiation. The cell undergoes several early divisions producing more totipotent cells, blastomeres that give rise to the embryonic membranes, placenta and the embryo. The self-renewing property of the stem cell is manifest by either symmetric or asymmetric cell divisions; symmetric divisions result in in the propagation of two stem cells or the production of two terminally differentiated cells, whereas asymmetric cell division results in one stem cell and one terminally differentiated cell. The cells have a hierachy of differentiation potential with the cells at the top of the heirachy derived from the first few cell divisions after fertilisation. Further down are the embryonic stem cells derived from the inner cell mass of the blastocyst. These pluripotent cells can differentiate into any of the three germ layers of ectoderm, mesoderm and endoderm. During embryonic development, stem cells from the blastocyst give rise to cell progenies that become progressively restricted in their phenotypic potential to generate mature tissue. At advanced stages of development, the product of stem cells may be a multipotent cell with limited degree of differentiation, lower self-renewal potential, and a higher cell proliferation rate. Adult somatic cells are usually terminally differentiated or have restricted phenotypes they can adopt under specific culture conditions. The cell divisions eventually produce terninally differentiated cells that are unable to renew and eventually undergo apoptosis (Alison et al, 2002; Triffitt, 2002). Many adult tissues maintain populations of cells that are not terminally differentiated. These postnatal stem cells are required for normal tissue remodelling and repair. They can be isolated from tissues of individuals of any age and maintain some capacity for multilineage differentiation.

EMBRYONIC STEM CELLS

The culture of embryonic stem cells derived from the blastocyst was first described by Thomson et al in 1998. These cells have generated considerable interest as a potential source for tissue engineering because of their high telomerase activity allowing them to proliferate indefinitely *in vitro,* and maintainence of their pleuripotency (de Wert and Mummery, 2003; Vats et al, 2005). Donor embryonic stem cells will however need to be histocompatibility leucocyte antigen (HLA)-matched to the recipient as a prerequeset to clinical application. Transplanted embryonic stem cells have caused teratomas in mouse models highlighting the need for using only fully differentiated cells in the clinical setting (Orkin and Morrison, 2002). These findings, as well as the potential in higher passage cells of epigenic and genetic changes resulting in an abnormal karyotype with trysin or collagenase IV (Draper et al, 2004), raise serious safety issues.

ADULT MESENCHYMAL STEM CELLS

Experiments performed by Friedenstein and co-workers described the presence of mesenchymal stem cells (MSCs) in the bone marrow (Friedenstein 1966; Friedenstein 1968). Friedenstein demonstrated that these cells could be isolated through their intrinsic property to adhere to tissue culture plastic (Friedenstein 1970). These cells formed colonies of cells with spindle-like fibroblastic appearance *in vitro*, and they were initially termed colony forming unit-fibroblasts (CFU-Fs). *In vitro* studies have shown that CFU-Fs are a heterogeneous population of stem cells at different levels of heirachy (Owen et al, 1987; Owen and Friedenstein, 1988). MSCs are cells derived from the mesoderm and are defined as cells that can give rise to a variety of mesenchyme derived cell types including chondrocytes, osteoblasts, adipocytes, myoblasts and hepatocytes (Prockop, 1997; Pittenger et al, 1999). These cells have considerable therapeutic potential for the repair and regeneration of tissue. By the end of last century, there was considerable interest in the use of MSCs for clinical tissue engineering applications highlighted by the work of Pittinger et al (1999) showing that cells could differentiate *in vitro* into the three mesenchymal lineages of chondrocytes, osteoblasts and adipocytes.

Stem cells possess self-renewal capacity, and exhibit long-term viability and multilineage differentiation potential. Ethical, political and religious issues surround the use of embryonic stem cells. In contrast, the use of autologous postnatal MSCs is generally well accepted by society. MSCs are less tumourogenic than their embryonic counterparts (Raghuath et al, 2005) and provide an autologous source of cells eliminating concerns regarding rejection and disease transmission. There is also evidence to suggest that MSCs have immunosuppressive potential as co-culture with MSCs inhibits T-cell lymphocyte proliferation (Krampera et al, 2003), and MSCs have also been shown to be negative for major histocompatability complex (MHC) class II antigens and the co-stimulatory molecules B7-1 and B7-2 (Devine and Hoffman, 2002; Majumdar et al, 2003) and they are being tested in clinical trials to treat GVHD (graft-versus-host-disease) arising from bone marrow allografts (Le Blanc et al, 2004) and also for the treatment of Crohn's disease (inflammatory bowel disease) and COPD (chronic obstructive pulmonary disease). All these factors support the use of MSCs for the creation of a more marketable off-the-shelf tissue engineered product.

Chapter 2

SOURCES OF ADULT MESENCHYMAL STEM CELLS

Cells with stem cell characteristics have been isolated from many different adult tissues including cord blood, peripheral blood, bone marrow, spleen, liver, kidney, thymus, dental pulp, periosteum, skin, retina, adipose tissue, skeletal muscle, synovial tissue and the synovial infrapatellar fat pad (Johnstone et al, 1998; Erices et al, 2000; De Bari et al, 2001; Zuk et al, 2001; Dragoo et al, 2003; Peng and Huard, 2003; Wickham et al, 2003). The choice of stem cell source is determined by ease of access to tissue source, frequency of stem cells and information on a particular cell system. More recent published work also suggests that MSCs from different tissues vary in their differentiation potential (Zuk et al, 2002; Sakaguchi et al, 2005).

BONE MARROW DERIVED MESENCHYMAL STEM CELLS

Bone marrow derived stem cells have been widely studied and there is a wealth of information in literature concerning them. Adult mammalian bone marrow contains two discrete stem cell populations, haematopoietic stem cells and MSCs (Pittinger et al, 1999; Short et al, 2003). Protocols for the culture (Freidenstein et al, 1970) and, chondrogenic, osteogenic and adipogenic differentiation of bone marrow derived MSCs have been described (Johnstone et al, 1998; Pittenger et al, 1999; Sekiya et al, 2002). However not all stromal cells derived from bone marrow are MSCs. These cells form only 0.001-0.01% of the total nucleated cells in bone marrow aspirates (Jones et al, 2002). Harvesting of bone marrow is painful with donor site morbidity and risk of wound infection and sepsis (Pittenger et al, 1999). Bone marrow aspirate of 30 ml only produces approximately 1×10^5 cells (Bruder et al, 1997a) making expansion in culture necessary. Obtaining a large number of cells at harvest has the potential advantage of not needing costly and time-consuming tissue culture expansion that risks cell contamination.

SYNOVIAL FAT PAD DERIVED STEM CELLS

The synovial infrapatellar fat pad or Hoffa's fat pad is an intracapsular but extrasynovial structure. It lies on the inferior aspect of the patella behind the patellar ligament and in front of the synovial lined knee joint (William et al, 1989). It is a flexible and displaceable structure. The volume varies and it is lost only in extreme emaciation after subcutaneous fat is eliminated (Saddik et al, 2004). The fat pad is similar in structure to subcutaneous adipose tissue containing fibrous tissue interspersed among adipose tissue. It may contain horizontal and vertical synovial lined clefts that communicate with the synovial cavity (Smillie et al, 1974; LaPrade, 1998). The synovial fat pad has a rich blood supply that is derived from an anastamosis of vertically orientated vessels from the superior and inferior genicular arteries. These vertical vessels are then further interconnected by horizontal vessels (Kohn et al, 1995; Kim et al, 1996).

MSCs extracted from synovial fat pad have been induced into chondrogenic, adipogenic and osteogenic phenotype using appropriate media (Dragoo et al, 2003; Wickham et al, 2003). Some papers describe the synovial fat pad as adipose synovium (Mochizuki et al, 2006), since the synovium covers the fat pad and the subsynovium comprises of adipose connective tissue. These cells have been shown to have a cell surface molecule profile similar but not identical to that of bone marrow derived MSCs, and maintain their multipotency into the later stages of life (Wickham et al, 2003).

Compared to bone marrow, synovial fat pad is reported to give a higher yield of adherent colony forming cells; bone marrow aspirate of 30 ml produced approximately 1×10^5 cells (Bruder et al, 1997a), whereas 21 ml of synovial fat pad yielded approximately 5.5×10^6 cells (Dragoo et al, 2003). There is reduced pain and morbidity associated with the harvest of synovial fat pad cells compared with bone marrow cells (Dragoo et al, 2003). In a patient matched quantitative comparison looking at MSCs from five sources, synovial tissue derived cells showed better proliferation and chondrogenic potential under the conditions tested compared to cells from bone marrow, adipose tissue, periosteum and skeletal muscle (Sakaguchi et al, 2005). They also showed better osteogenic potential, along with bone marrow and periosteum derived cells, and better adipogenic potential along with adipose tissue derived stem cells. In another study, fat pad derived cells were shown to be more similar in their cell surface epitope profile, and proliferative, chondrogenic and osteogenic differentiation potential to synovial tissue derived cells than to adipose tissue derived cells (Mochizuki et al, 2006).

Synovial fat pad derived MSCs are a possible alternative to differentiated chondrocytes in autologous chondrocyte implantation for the repair of focal cartilage defects. Compared to cells harvested from the bone marrow, these cells are easier to obtain, have lower donor site morbidity and are associated with a higher yield of MSCs (Dragoo et al, 2003). A biopsy from these tissues may represent an easily accessible source of mesenchymal stem cells at diagnostic or therapeutic arthroscopy. The synovial fat pad is commonly resected at arthroscopy and total knee arthroplasty for improved surgical visualisation, and in arthroplasty to prevent possible impingement of the fat by the prosthesis. No adverse long-term side effects have been noted following the resection of the synovial fat pad (Duri et al, 1996). It is also resected for chronic impingement and fibrosis of the fat pad (Hoffa's disease) (Ogilvie-Harris and Giddens, 1994). To enable their use in cell-based repair strategies, it is crucial that more is known about the exact nature of these cells.

AGE-RELATED CHANGES IN MESENCHYMAL STEM CELLS

The reported effects of ageing on MSCs are variable. A number of studies have shown an age-related decline in the number and proliferation of bone marrow derived MSCs (D'Ippolito et al, 1999; Shamsul et al, 2004; Bertram et al, 2005; Huang et al, 2005a; Mareschi et al, 2006; Stolzing et al, 2008; Zhou et al, 2008), however a number of other studies found no difference (Stenderup et al, 2003; Suva et al, 2004; Scharstuhl et al, 2007). Our preliminary experiments on bone marrow derived MSCs showed no significant age-related changes in cell proliferation and characterisation supporting the work of Mareschi et al (2006). A number of studies have shown an age-related change in the differentiation potential of bone marrow derived MSCs (Jiang et al, 2008; Zhou et al, 2008), but this has not been reported in some other studies (Stenderup et al, 2003; Huang et al, 2005a; Roura et al, 2006; Scharstuhl et al, 2007; Siddappa et al, 2007). Scipper et al (2008) compared adipose tissue derived MSCs and noted differences in cell proliferation and PPAR-gamma-2 expression related to age. The applicants have previously looked at the effects of ageing in later life (age 50-90 years) on the isolation, expansion, cell surface characterisation and osteogenic potential of synovial fat pad derived MSCs and found no difference (Khan et al, 2009). Much of the conflicting data in the literature may be due to variations between patients and in the culture conditions used. In this proposed study, the effect of such variations will be reduced via examinations of the effect of age on the phenotype, proliferation and differentiation potential of patient-matched samples during identical culture conditions.

A potential cell source that does not show age-related decline in proliferation and differentiation is important in determining the optimal cell-based tissue repair therapies in an aging population. An ideal source of stem cells would be easy to obtain with a small risk of complications, with a good cell yield not requiring long culture expansion and exhibit good proliferation and differentiation potential.

Chapter 3

DIFFERENTIATION OF STEM CELLS

MSCs have been shown to differentiate into chondrocytes, osteoblasts and adipocytes under appropriate culture conditions (Pittenger et al, 1999). Predifferentiation of MSCs will be essential in clinical applications to ensure appropriate lineage commitment and to avoid undesired tissue formation and heterotopic tissue formation. The presence of only the *in vivo* environment alone is not sufficient to allow chondrogenesis (Wakitani et al, 1994; Manne et al, 2005). Chondrogenesis has been shown to occur in MSCs cultured as cell aggregates with specific differentiation and growth factors (Johnstone et al, 1998; Mackay et al, 1998). Chondrogenic differentiation of such cells requires specific medium with TGF β and a three-dimensional culture environment e.g. as a cell aggregate, to allow differentiation *in vitro*. This shows that cell density, contact and topography also contribute to the process (Johnstone et al, 1998; Pittenger et al, 1999; Sekiya et al, 2002). Mackay et al (1998) showed that high glucose medium and TGF β3 are needed for the chondrogenic differentiation of MSCs. Insulin-like growth factor-1 (IGF1) has been shown to have a synergistic effect with TGF β in promoting chondrogenesis in MSCs (Indrawattana et al, 2004). The end point is the accumulation of cartilage matrix gene products as demonstrated by gene expression and immunostaining of collagen types II, IX and XI, and aggrecan.

Chondrocyte differentiation can be divided into three stages: cell condensation, chondrocyte differentiation and chondrocyte hypertrophic maturation. TGF β has a role in cell condensation, probably by inducing the expression of Sry-related HMG box-9 (SOX9) (Furumatsu et al, 2005). SOX9 regulates the expression of aggrecan and collagen types II, IX and XI during chondrocyte differentiation (Magne et al, 2005a). SOX5 and SOX6 are also expressed early during chondrocyte differentiation, but their exact molecular mechanisms remain uncertain (Akiyama et al, 2002). Hypoxia inducible transcription factor-1 (HIF1) makes chondrocyte survival possible in hypoxic conditions (Schipani et al, 2001). Hypertrophic differentiation is characterized by the expression of collagen type X.

Adipogenic and osteogenic differentiation can also be demonstrated with the same cell population (Pittenger et al, 1999), but the detailed molecular mechanisms driving differentiation into these different cell phenotypes have not yet been elucidated. Monolayer culture of mesenchymal stem cells with standard osteogenic medium containing β-glycerophosphate, dexamethasone and ascorbate have been shown to induce osteogenic differentiation (Beresford et al, 1994). Adipogenesis in undifferentiated MSCs is comprised of two stages: determination of the adipocyte lineage and adipogenic differentiation. The molecular pathways involved in the second stage have been studied extensively (Gregoire, 2001) but those for the first stage are not well understood (MacDougald and Mandrup, 2002). It has previously been reported that the adipogenic differentiation of MSCs is only possible when the cells are confluent (Ishino et al, 2004). MSCs require several cycles of hormonal stimulation to exit from the cell cycle: a feature necessary to achieve commitment to the adipogenic lineage (Ramirez-Zacarias et al, 1992).

Insulin, dexamethasone and 3-isobutyl-1-methyl xanthine (IBMX) have been shown to be sufficient to stimulate adipogenic differentiation (Pittinger et al, 1999; Zuk et al, 2001; Rim et al, 2005). Insulin is needed to generate the substrate glycerol 3-phosphate, which is needed for the biosynthesis of triglycerides (Sottile and Seuwen, 2001). Insulin is known to promote the proliferation and differentiation of pre-adipocytes (Ailhaud, 1982). High concentrations of insulin mimic the role of IGF1 (Qiu et al, 2001) and have a mitogenic effect. The anti-inflammatory drug indomethacin is a peroxisome proliferator-activated receptor gamma-2 (PPARγ2) activator and is a strong inducer of adipogenesis (Rosen and Spiegelman, 2000).

OPTIMISATON OF CHONDROGENIC DIFFERENTIATION POTENTIAL

Stem cells are defined by their self-renewal and multipotentiality. Unfortunately these crucial properties appear to show considerable donor variability and become limited on expansion in monolayer culture as these cells lose their proliferation and differentiation potential (Pittinger et al, 2001;

Cancedda et al, 2003). As expansion in culture is needed to increase the cell number to a level suitable for clinical applications, it is important to explore avenues that would allow for this expansion without a significant compromise of differentiation potential.

Cells are normally cultured *in vitro* in an atmosphere of 5% carbon dioxide in standard tissue culture incubators. As the remaining gas consists of air, oxygen levels are almost 20% (140 mm Hg). It has long been known that some cells, including some with stem cell characteristics, proliferate more rapidly in lower oxygen concentrations (Rich and Kubanek, 1982; Rich, 1986). These cells include human periosteal cells (Deren et al, 1990), haematopoietic progenitor cells (Koller et al, 1992) and neural crest stem cells (Morrison et al, 2000). This would appear feasible since the partial pressure in arterial blood is 75-110 mm Hg and that in bone marrow is 27-49 mm Hg (Lennon et al, 2001). Culture in low oxygen selectively enhances collagen type II expression and cartilage matrix assembly (Adesida et al, 2006), and appears to be a major and selective control at the translational level with upregulation of collagen processing enzymes.

Articular cartilage is avascular and exists at reduced oxygen tension of 1-7% in vivo depending on the depth from the articular surface (Silver, 1975; Wang et al, 2005). So, it is not surprising that hypoxia is also known to increase the synthesis of extracellular matrix components of chondrocytes (Murphy and Sambanis, 2001; Domm et al, 2002). In stem cells derived from bone marrow (Scherer et al, 2004), adipose tissue (Wang et al, 2005) and synovial fat pad (Khan et al, 2007), hypoxia has also been shown to improve chondrogenesis.

Monolayer expansion is associated with a flattened cell morphology in stem cells and in primary chondrocytes. Chondrocytes have been shown to maintain their chondrogenic phenotype when cultured under conditions that prevent cell flattening e.g. high-density cell aggregate culture (Watt, 1988) or three-dimensional scaffolds, and by supplementation of actin disrupting agents (Loty et al, 1995). FGF-2 has been shown to induce disassembly of the actin microfilament architecture (Wroblewski and Edwall-Arvidsson, 1995). Disruption of actin microfilaments results in the upregulation of SOX9 mRNA (Tew and Hardingham, 2006), and this may be the pathway involved in maintaining the chondrogenic phenotype.

FGF-2 is also a potent mitogen for a variety of cell types derived from the mesoderm including chondrocytes (Kato and Gospodarowicz, 1985). The potential of passaged articular chondrocytes to be chondrogenic has been shown to be enhanced by rapid expansion with a medium containing PDGF-

BB, TGF β3 and FGF-2 (Hardingham et al, 2002; Martin et al, 2003; Li et al 2004). FGF-2 has been shown to enhance proliferation and differentiation of bone marrow (Martin et al, 1997; Bianchi et al, 2003; Solchaga et al, 2005) and synovial fat pad (Khan et al, 2008) derived MSCs. FGF also has a role in *in vivo* cartilage repair when released from bone matrix where it is stored, and following local exogenous administration into osteochondral defect cavities (Hiraki et al, 2001).

Chapter 4

STEM CELL MARKERS

Most cell surface markers are inadequate in identifying stem cells unambiguously either because these markers are also expressed by non-stem cells, or they are only expressed by stem cells at a particular stage and under particular culture conditions (Pittinger and Martin, 2004). Tables 1 and 2 shows known cell markers for bone marrow derived haematopoietic stem cells and MSCs, and synovial fat pad derived MSCs. Bone marrow MSCs are uniformly positive for CD29, CD44, CD71, CD90, CD105 and CD106, and are negative for markers of haematopoetic lineage including CD14, CD34 and CD45 (Haynesworth et al, 1992; Barry et al, 1999; Pittenger et al, 1999; Jones et al, 2002). Chondrogenic potential appears to be related to the expression of CD105, a TGF β receptor, recognised by SH2 (Barry et al, 1999; Ragunath et al, 2005).

Bone marrow derived MSCs are selected from fresh aspirate as STRO1 and CD106 positive (Dennis et al, 2002; Shi and Gronthos et al, 2003). CD34 contains many cell types in addition to early progenitor and stem cells, and STRO1 antibody was developed to differentiate MSCs from other CD34 positive cells in the bone marrow (Simmons and Torok Stork, 1991; Gronthos et al, 1994; Gronthos et al, 1999). STRO1 does not bind to haematopoietic progenitor cells but binds to bone marrow derived MSCs (Dennis and Caplan, 2000; Dennis et al, 2002). It identifies an as yet uncharacterised cell surface molecule.

Alpha smooth muscle actin (αSMA) is also expressed by freshly isolated bone marrow derived MSCs and 70% of the STRO1 positive cells express αSMA (Shi and Gronthos, 2003).

Table 1. Cell markers used in the characterization of mesenchymal stem cells and their characteristics

Cell Marker	Characteristics
3G5	Monoclonal antibody that recognises CD44v3
STRO-1	Murine IgM monoclonal antibody that identifies a trypsin insensitive unidentified cell surface antigen on bone marrow derived MSCs
αSMA	Alpha smooth muscle actin, specific for smooth muscle fibres
αLNGFR	Receptor for low affinity nerve growth factor
SH2	Monoclonal antibody that identifies CD105
SH3	Antibody that identifies CD73
SH4	Antibody that identifies CD73
SB-10	Monoclonal antibody that recognises CD166
CD10	Common acute lymphocytic leukaemia antigen
CD11b	Integrin αM subunit
CD13	Aminopeptidase N
CD14	Component of lipopolysaccharide receptor on lymphocytes
CD29	Integrin β1 subunit
CD34	Cell surface glycoprotein of haematopoietic lineage
CD44	Receptor for hyaluronan
CD44v3	Receptor for a ganglioside on pericytes in the vasculature
CD45	Leucocyte common antigen, a protein tyrosine phosphatase
CD49a	Integrin alpha subunit
CD56	Neural cell adhesion molecule (NCAM)
CD73	GPI-linked cell surface protein involved in B cell activation
CD90	GPI-linked cell surface protein Thy-1
CD105	Endoglin; TGF β receptor type III
CD106	Vascular cell adhesion molecule 1 (VCAM-1)
CD146	MCAM/Muc-18, antigen on endothelail cells
CD166	ALCAM, Type I membrane glycoprotein adhesion molecule on activated leucocytes
D7FIB	Marker of human fibroblast/epithelial cells
vWF	von Willebrand factor: product of endothelial cells/platelets

Table 2. Cell markers for bone marrow derived haematopoietic stem cells and MSCs in aspirate and culture, and for synovial fat pad derived MSCs

Tissue	Cell markers
Bone marrow derived haematopoietic stem cells	CD14, CD34, CD45 (Haynesworth et al, 1992) CD11b, CD34, CD45 (Baddoo et al, 2003)
Bone marrow derived MSCs in aspirate	αSMA (Shi and Granthos, 2003) αLNGFR (Quirici et al, 2002) STRO1, CD106 (Simmons and Torok Stork, 1991; Dennis et al, 2002; Shi and Gronthos, 2003) LNGFR, STRO1, CD10, CD13, CD90, CD105 (Jones et al, 2002)
Bone marrow derived MSCs in culture	CD29, 44, 106 (Baddoo et al, 2003) αSMA (Shi and Granthos, 2003) SH2, SH3, SH4, CD44, CD90 (Pittenger et al, 1999) SH2, SH3, SH4 (Haynesworth et al, 1992)
Synovial fat pad derived MSCs	CD13, CD29, CD44, CD59, CD105 (Wickham et al, 2003) CD 44, CD90, CD105, CD147 (Mochizuki et al, 2006)

Monoclonal antibodies to the alpha low affinity nerve growth factor receptor (αLNGFR) stain freshly isolated bone marrow MSCs but do not label haematopoietic cells (Quirici et al, 2002). Baddoo et al (2003) used a combination of plastic adherence and in vitro culture along with the removal of contaminating haematopoietic cells by negative selection using antibodies to CD11b, CD34 and CD45. The resulting cells were shown to express CD29, CD44 and CD106. This describes phenotypical properties of bone marrow MSCs based on analysis of marrow stromal cells in culture. Jones et al (2002) reported these cells to be uniformly positive for LNGFR, STRO1, CD10, CD13, CD90 and CD105, and negative for CD14, CD34, CD117 and CD133.

Wickham et al (2003) showed that more than 50% of the synovial fat pad derived cells in culture were positive for CD13, CD29, CD44, CD59 and CD105. Mochizuki et al (2006) showed that the same proportion of cells were positive for CD44, CD90, CD105 and CD147. Other potential MSC markers for culture-expanded MSCs include CD10, CD31, CD49a, CD54, CD55,

CD73, CD146 and CD166 (Vaananen, 2005). Antibody SB-10 reacts with CD166 (Bruder et al, 1998) and SH3 and SH4 identify CD73 (Barry et al, 2001). We have shown that all bone marrow derived MSCs (unpublished data) and fat pad derived MSCs (Khan et al, 2007) stain strongly for CD13, CD29, CD44, CD90 and CD105.

A full characterisation of synovial fat pad derived MSCs is important to achieve a greater understanding of their origin and their repair potential. *Ex vivo* expanded cells may need to undergo a large number of cell divisions in monolayer culture to reach a number sufficient for clinical applications. It has previously been reported that MSCs retain their pleuripotency for 6-10 passages (Spees et al, 2004). De Bari et al (2001) showed that the cell surface epitope profile of synovial tissue derived MSCs was stable during expansion from passage 3 up to at least passage 10. We have shown that the cell surface characterisation of mesenchymal stem cells derived from the synovial fat pad is maintained with expansion in FGF-2 (Khan et al, 2008), with ageing in later life and (Khan et al, 2009) and with expansion up to passage 18 (unpublished data).

PROBLEMS WITH STEM CELL MARKERS

There is a lack of specific cell markers for selectively isolating stem cells. There are also potential problems with the use of cell surface markers including some variability in expression dependant on culture conditions (Bruder et al, 1997b; Stewart et al, 1999; Quirici et al, 2002).

STRO1 cross-reacts with erythroblasts. CD10, CD13 and CD90 are expressed on human fibroblasts as well as bone marrow derived MSCs (Jones et al, 2002). CD105 is also expressed on endothelial cells and early B lineage precursor cells in bone marrow (Barry et al, 1999; Deans and Moseley, 2000). CD44 and CD29 have broad cell reactivity (Jones et al, 2002; Quirici et al, 2002). Although CD34 is a haematopoetic stem cell marker, it is also expressed by endothelial cells (Miranville et al, 2004). Some studies report the expression of CD34 in MSCs isolated from the bone marrow, though they state that it is rapidly lost after in vitro culture (Quirici et al, 2002). The expressions for STRO-1 (Bruder et al, 1997b; Stewart et al, 1999) and LNGFR (Quirici et al, 2002) are also progressively lost in culture. This may be because of the progressive loss in culture of non-proliferating cells or presence of high proportions of foetal calf serum in the culture medium inhibiting the expression of some surface antigens (Garcia-Pacheco et al, 2001).

The use of a large number of markers and the negative selection procedures are used to deal with this problem. Pittinger et al (1999) reported the use of several antibodies to characterise expanded MSC populations from bone marrow.

PERICYTES AS CANDIDATE STEM CELLS

It has been suggested that bone marrow derived MSCs originate from microvascular pericytes (Bianco et al, 2001; Short et al, 2003). The exact nature and location of stem cells in the synovial fat pad, or indeed any tissue, is not known.

The inner layer comprising of a single layer of longitudinally arranged flattened epithelial cells is called the endothelium. This is supported by a basement membrane, collagenous tissue and an internal elastic lamina in arteries. This together with endothelium forms the tunica intima. The tunica media is the intermediate muscular layer and contains circumferentially arranged vascular smooth muscle cells. The tunica adventitia is the outer supporting tissue layer that merges with the surrounding collagenous tissue and may contain the external elastic lamina. Fibroblasts are the predominant cell type in this layer.

In arteries the amount of elastic tissue decreases with size and there is an increase of the smooth muscle content. Arterioles are vessels with a lumen of less than 0.3 mm in diameter. They have a thin internal elastic lamina and no external elastic lamina. The tunica media is composed of smooth muscle cells in six concentric layers or less. The tunica adventitia may be as thick as the tunica media. In venules, the elastic and muscular components are much less prominent. A characteristic feature is the wide lumen diameter relative to the wall thickness. Muscular venules have an intimal layer with no elastic fibres and a medial layer consisting of one or two layers of smooth muscle fibres. Capillaries consist of a single layer of flattened endothelial cells lining the capillary lumen. Muscular and adventitial layers are absent. Pericytes in the form of flattened cells embrace the endothelial cells.

The term 'pericyte' originates from its early anatomical descriptions ('peri-' around and 'cyto-' cell) reflecting its periendothelial location of these cells. Pericytes are cells closely associated with capillaries and were first described by Rouget in 1873 (Hirschi and D'Amore, 1996). Vascular pericytes are cells that were initially characterised as associated with retinal capillaries. The embryonic origin of pericytes remains unknown but it is suggested that they may be derived from the mesenchyme (Nayak et al, 1988). They are embedded within a basement membrane and are separated from the endothelial cells, that they surround, by the basal lamina that allows interdigitation (Nayak et al, 1988). They cover microvasculature arterioles, capillaries and venules on their abluminal surface. Pericytes are thought to be a heterogeneous cell population exhibiting tissue-related and vessel-related characteristics (Hirschi and D'Amore, 1996). Within the blood vessel, pericytes have a contractile function and are thought to control local blood flow (Hirschi and D'Amore, 1996). They also play an important role in microvessel stability, structural integrity, vasodynamic capacity, permeability and control of angioneogenesis (Hirschi and D'Amore, 1997; Thomas, 1999).

The cell biology of pericytes was largely developed by investigation of retinal peicytes isolated from bovine eyes and these were of particular interest as potential sources of vascular calcification. Pericytes isolated from bovine retinal capillaries were reported to be STRO1 positive (Doherty et al, 1998) and exhibited potential for differentiation into a variety of cell types including osteoblasts (Brighton et al, 1992), smooth muscle cells (Meyrick et al, 1981), adipocytes and chondrocytes (Rhodin, 1968; Farrington-Rock et al, 2004). Pericytes have also been reported in other tissues such as dental pulp as antibody 3G5 positive cells (Bankfalvi et al, 1998; Nayak et al, 1988), and in bone marrow as STRO1 positive cells (Doherty et al, 1998; Shi and Gronthos, 2003). Pericytes, like bone marrow MSCs, express STRO1, αLNGFR and αSMA in culture (Herman and D'Amore, 1985; Wesseling et al, 1995; Jones et al, 2004).

αSMA is also expressed by freshly isolated bone marrow MSCs *in vitro*, and 70% of the STRO1 positive cells express αSMA (Shi and Granthos, 2003). αSMA has restricted distribution in bone marrow *in vivo* in the vascular smooth muscle cells in the tunica media of the arteries and pericytes, and occasional flattened cells on the endosteal surface of bone (Bianco et al, 2001).

The periosteal pericytes are in the correct anatomical location for migration into the bone marrow during development. It has been suggested that if distributed more widely with capillaries, pericytes could account for stem cells in other tissues (Gronthos et al, 2003). In support of this theory, a continuous subendothelial network of pericyte like cells has been identified using 3G5 throughout the entire human vascular bed (Andreeva et al, 1998). Indeed, many of the tissues from which stem cells have been isolated have good vascularisation.

A minor population of bone marrow derived MSCs has been found to be positive for the cell surface ganglioside recognised by antibody 3G5 (Shi and Gronthos, 2003; Khan et al, in press).

Chapter 5

ARTICULAR CARTILAGE

Articular or hyaline cartilage is a specialised connective tissue that is avascular, aneural and alymphatic. It is a load bearing tissue supported by underlying subchondral bone. Chondrocytes are the only cell type and there is a low cell turnover. These cells are highly specialised cells that secrete the extracellular matrix proteins. They function by maintaining the integrity of the cartilage by balanced synthetic and catabolic activities. The extracellular matrix is composed of a complex combination of predominantly type II and lesser amounts of type VI, IX and XI collagen fibrils specifically arranged with large water retaining aggrecan molecules and other smaller proteoglycans bonded to them. This combination gives cartilage the ability to resist repetitive compressive load bearing without premature wear (Poole, 1995; Hardingham, 1998).

Collagens contain a unique triple-helical peptide structure with a repeating motif, with glycine as every third amino acid and a high frequency of proline. These molecules are arranged in fibrils with overlapping and cross-linking of adjacent molecules. The main collagen in cartilage is collagen type II forming 80-90% of the total content. It is a long chain fibrillar collagen that forms the major fibre network of the tissue. Collagen type XI is also a long chain fibrillar collagen and accounts for almost 3% of the total collagen content in cartilage. Collagen type IX is found on the outside of type II fibrils, and has a special structure possibly to allow interactions with other fibrils and proteoglycans. Collagen type VI may have a role in cell-matrix interactions. Collagen type X is a short chain collagen that occurs specifically in calcifying cartilage produced by hypertrophic chondrocytes in the growth plate (Hardingham, 1998).

The interfibrillar matrix is filled with proteoglycan. The main proteoglycan in cartilage is aggrecan that consists of a large core protein with polysaccharide chains attached to it including chondroitin sulphate and keratan sulphate. At the N-terminal of the protein core there are two globular domains G1 and G2. The G1 domain acts as a site for aggregation of aggrecan molecules, and allows binding to hyaluronan and link protein (Hardingham, 1998). Almost 70% of the cartilage wet weight is water and this contributes to the load-bearing properties of the tissue. The glycosoaminoglycan (GAG) chains attached to proteoglycans carry a high density of negative charge. This attracts positively charged ions and creates an osmotic pressure that draws water into the tissue (Hardingham and Fosang, 1992).

EMBRYOLOGY

Articulating joints start forming embryologically within cartilage long bone rudiments with the appearance of regions of high cell density, interzones, where cells lose the expression of chondrocyte-specific markers such as collagen type II. These cells differentiate and form three layers. The central layer has a lower cell density and the cells die through apoptosis creating a joint cavity. Cells on either side differentiate to form articular chondrocytes. Adult articular cartilage can also be divided into three layers. The surface zone is characterised by flattened, discoid cells that mainly secrete proteoglycan. The mid zone contains rounded cells arranged in columns that mainly secrete collagen type II and aggrecan. The deeper zone is calcified and contains collagen type X (Eyre, 1991; Magne et al, 2005b).

GROWTH PLATE CARTILAGE

Bone is formed through two separate developmental processes where mesenchymal cells are recruited into condensations before differentiation on receiving signals elicited by factors such as members of the transforming growth factor beta (TGFβ) superfamily. In intramembranous ossification, the cells directly commit to become osteoblasts, but in endochondral ossification, the cells first form a cartilaginous template in the form of the growth plate. The cells in the centre of the condensations differentiate into chondrocytes promoted by strong SOX9 activation and low Wnt/β-catenin signalling. Later in development, the growth plate chondrocytes become flattened and organised into columns. These chondrocytes express early markers such as collagen type II, IX and XI, and aggrecan. The cells on the peripheries of the condensation form the perichondrium and differentiate into osteoblasts.

The chondrocytes then further differentiate to become prehypertrophic chondrocytes and then hypertrophic chondrocytes that express collagen type X and calcify their extracellular matrix. The hypertrophic maturation of the chondrocytes is under the control of the Ihh/PTHrP loop; Ihh expressed in prehypertrophic cells induces the expression of PTHrP in the cells located in the perichondrium. PTHrP in turn inhibits hypertrophic differentiation and failure of its signalling results in Blomstrand chondroplasia. Members of the Wnt family on the other hand trigger hypertrophic differentiation. These hypertrophic chondrocytes eventually die through apoptosis and the vascular invasion of the calcified cartilage brings osteoclasts and osteoblasts (Karsenty and Wagner, 2002; Magne et al, 2005b).

OSTEOARTHRITIS

Articular cartilage is frequently damaged by trauma and in joint disease, such as osteoarthritis but shows only a limited capacity for repair. This is due to the lack of inherent mechanisms of repair in mature articular cartilage. Cartilage defects that extend to the subchondral bone show some signs of repair with the formation of neocartilage (Newman, 1998) probably due to the release of bone marrow derived stem cells from the underlying subchondral bone (Shapiro et al, 1993).

Osteoarthritis is the most prevalent disorder of the musculoskeletal system. Approximately fifteen percent of the total UK population suffers from arthritis, and seventy percent of the population over the age of 65 years will limit their activities or seek medical attention because of osteoarthritis. It has a significant impact on the ability to perform the activities of daily living (Simon, 1999). This together with the increasing age of the population and the costs involved in management of this disorder make it a major social issue, especially in industrialised developed countries with high life expectancy (Buckwalter, 2002). A total knee or hip replacement costs the National Health Service (NHS) between four and eight thousand pounds and the additional cost to the individual in terms of socio-economic consequences with loss of earnings and productively, and social and psychological upset is significantly higher. This is the driving force behind numerous ongoing efforts to develop new tissue-engineered strategies for the treatment of osteoarthritis (Schulz and Bader, 2007).

PATHOPHYSIOLOGY OF OSTEOARTHRITIS

The main pathological features of osteoarthritis, including cartilage failure, are characterised by dysregulation of tissue turnover in articular cartilage and subchondral bone. This leads to an imbalance between dynamic reparative and catabolic processes (Hardingham, 1998), where both increased cartilage matrix turnover and increased production are inviolved. The tissue damage may be driven by local production of inflammatory cytokines e.g. tissue necrosis factor alpha (TNF α) and interleukin-1 beta (IL-1β) and protease release by cells in cartilage, synovium and bone (Risbud and Sittinger, 2002) and the anabolic response in cartilage may involve local release of FGF2 (Chia et al, 2009), but may also be contributed to by TGFβs, bone morphogenic proteins (BMPs) and insulin-like growth factor-1 (IGF1). This leads to a loss of proteoglycan and disruption of the collagenous fibrillar network. While osteochondral defects show some evidence of attempts at repair with the formation of neocartilage (Newman, 1998), chondral defects do not. This is probably due to the release of bone marrow derived stem cells and growth factors in cartilage defects that extend to the underlying subchondral bone (Shapiro et al, 1993; Newman, 1998).

The patients frequently complain of pain, stiffness and swelling of the joints that leads to reduced exercise tolerance and poorer quality of life. The first line of management is a trial of non-operative treatment that includes analgesia or anti-inflammatory medication, physiotherapy and occupational therapy. The articular cartilage shows only a limited capacity for repair and current treatments are aimed at relieving inflammation and pain and do little to delay disease progression (Wollheim, 1996; Simon, 1999).

Chapter 6

CURRENT SURGICAL TREATMENT MODALITIES

Most patients with osteoarthritis end up requiring surgical intervention. A few options for the repair of focal cartilage lesions have been used in larger joints including abrasive chondroplasty, subchondral drilling, microfracture and cartilage transplantation. In subchondral drilling and microfracture, bone marrow derived stem cells are stimulated to migrate from the subchondral bone to the site of cartilage defect. This however results in the formation of fibrocartilage rather than hyaline cartilage. Histologically, fibrocartilage contains more fibrous tissue, and biochemically contains significantly less proteoglycan and more collagen type I. Fibrocartilage has inferior mechanical and hydroelastic characteristics and results in unsatisfactory clinical outcome (Hunziker, 2001). Cartilage transplantation is limited by donor availability and is only available in a few centres with variable results (O'Driscoll, 1998; Bentley and Minas, 2000; Hunziker, 2001). Most focal cartilage lesions, left untreated, progress to more extensive lesions and these require extensive surgical procedures in the form of joint arthroplasty or arthrodesis.

The frequent outcome when pain and loss of function become severe is surgical intervention for joint replacement or arthrodesis. Joint replacement is successful for elderly patients, but the limited lifetime of prostheses makes it much less desirable for younger patients. Younger patients are more likely to return to a more active lifestyle putting increased stresses and strains on their joint replacement (Parsons and Sonnabend, 2004). This means a greater likelihood of needing a revision procedure with its associated increased operative complications.

AUTOLOGOUS CHONDROCYTE IMPLANTATION

For joint problems in younger patients there is great interest in the potential of cell-based strategies to provide a biological repair of cartilage. Autologous chondrocytes are being used for the repair of focal cartilage defects in larger joints combined with a periosteal or resorbable collagen membrane sutured to the cartilage surface (Brittberg et al, 1994, 2001, 2003). These principles could also apply as proof-of-principle to focal defects in the hand. Although this technique was initially described for the knee, it has now been extended to the hip, ankle, shoulder, elbow and wrist joint (Brittberg et al, 2003). This is becoming a generally accepted technique for the repair of focal defects in the articular cartilage and has proved superior to mosaicoplasty in a prospective controlled clinical trial (Bentley et al, 2003). The first stage of the procedure entails an initial biopsy from a low weight bearing part of the articular cartilage. This is digested and mature articular chondrocytes are retrieved. These are then expanded in monolayer culture *ex vivo* for three weeks. The proliferation allows the number of chondrocytes to increase. Chondrogenesis is confirmed by gene expression for collagen type II and aggrecan, and histochemical and biochemical analysis. The cells are then trypsinised and ready for reimplantation at the second stage. The second stage involves an arthrotomy, debridement of articular defect and suturing of a periosteal or resorbable collagen membrane sealed with fibrin glue. The cells are then delivered under the membrane and the wound is closed. The cells settle, adhere, proliferate and lay down new cartilage (Brittberg et al, 2003).

The procedure has demonstrated usefulness, but would benefit from further development. It is a two-step procedure, it is expensive and technique-dependant. The procedure involves cartilage biopsy from the joint to obtain the chondrocytes and the periosteum, which therefore involves injury to the low weight bearing cartilage surface at initial biopsy and possible injury to the periosteum at the second procedure with associated morbidity of the donor site. It has only limited applications i.e. for focal cartilage defects, in view of the amount of tissue that can be obtained. Human articular chondrocytes are not easily extractable. The small amount of tissue also means that cell expansion in culture is necessary, and this is generally slow (Homicz et al, 2002). With prolonged expansion chondrocytes lose their ability to proliferate (Risbud and Sittinger, 2002; Cancedda et al, 2003) and their ability to express cartilage specific proteins (Benya and Shaffer, 1982). Also, the number of cell divisions these cells can undergo reduces with age and deteriorating health restricting the expansion potential of these cells (Dozin et al, 2002). These cells, when implanted, result in the formation of fibrocartilage rather than the desired hyaline cartilage (Clar et al, 2005). Although short-term clinical results have been good, evidence suggests progression of degenerative changes in the joint (Lee et al, 2000).

The current autologous chondrocyte procedure thus implants cells with variable chondrogenic potential. It is thus less than ideal to use autologous chondrocytes harvested from the same joint.

Chapter 7

TISSUE ENGINEERING APPROACHES FOR ARTICULAR CARTILAGE DEFECTS

Articular cartilage is a particularly suitable tissue for tissue engineering applications as it is avascular, aneural and alymphatic. Articular cartilage shows a limited capacity for repair following injury. Cartilage injuries that extend down to the subchondral bone show some signs of repair due to the release of bone marrow derived mesenchymal stem cells from the subchondral bone, a principle used in the surgical procedure of microfracture. More recently there has been an interest in cell-based strategies for the repair of articular cartilage including autologous chondrocyte implantation (ACI) (Peterson et al 2003), but the results are variable (Lee et al 2000).

A potential alternative source to the use of primary chondrocytes for cell-based cartilage repair strategies, that has been a focus of recent attention, is the use of stem cells or undifferentiated adult progenitor cells. Stem cells possess self-renewal capacity, and exhibit long-term viability and multilineage differentiation potential. In the first instance the alternative stem cell source will potentially be used to replace mature chondrocytes as a cell source in cell-based cartilage repair strategies. The use of stem cells should increase the consistency and reproducibility of the autologous chondrocyte implantation procedure. This could, in the future, be extended to treat more generalised cartilage defects, especially if the cell source is not a limiting factor unlike mature chondrocytes. Cartilage, being an avascular tissue, is an ideal target for tissue engineering.

Tissue engineering applications using MSCs present an interesting and promising new approach for the repair of articular cartilage defects (Hardingham et al, 2002).

Mesenchymal stem cells have been shown to differentiate into chondrocytes and represent an alternative cell source for therapeutic repair. Animal studies have successfully reported on the use of bone marrow derived mesenchymal stem cells embedded in a collagen gel to repair chondral defects (Wakitani et al 1994). To date there have been only limited reports of human autologous bone marrow derived cell implantation for cartilage repair (Wakitani et al, 2002; Wakitani, 2005; Kuroda et al, 2007) where expanded cells were embedded within a collagen scaffold, to repair a full-thickness cartilage defect in the knee. In a clinical trial, bone marrow derived mesenchymal stem cells injected into the medial femoral condylar cartilage defects at the time of high tibial osteotomy in twelve patients resulted in clinical improvement compared to the control group where no cells were used (Wakitani et al 2002). Although the score for clinical improvement was not significantly different, the patients treated with bone marrow derived MSCs did have better arthroscopic and histological grading scores. In another study, histological studies suggested that the defect was filled with hyaline-like type of cartilage tissue that stained positively with Safranin-O (Kuroda et al, 2007).

The optimal conditions for chondrogenic differentiation in MSCs derived from various sources are still being developed. We have shown that chondrogenesis in fat pad derived MSCs can be optimised by expanding cells in FGF-2 (Khan et al 2008) and allowing them to differentiate under hypoxic culture conditions (Khan et al 2007).

The management of cartilage defects is currently suboptimal and stem cells have the potential to improve on the current treatment modalities by replacing damaged tissue with hyaline cartilage. Tissue engineering holds promise for the future, but a number of challenges need to be overcome. These include further work on identifying the nature of stem cells and better characterisation, identifying the optimal source of stem cells for cartilage repair, and also identifying conditions that enhance expansion and chondrogenesis. Tissue engineering applications for the repair of cartilage rely on cells, scaffolds and signalling molecules, either in isolation or in combination.

The mode of delivery of cells for the repair of articular cartilage will depend on the size of the defect. Smaller well-localised defects can potentially be repaired by the direct injection of cells in the defect similar to current ACI procedures, whereas larger defects would invariably have to rely on an appropriate scaffold similar to current MACI procedures and, with increasing size, also on appropriate signalling molecules.

Stem cells could also potentially be used as a vehicle for gene delivery by transfecting cells with recombinant DNA constructs and coding for the expression of certain proteins and growth factors that promote chondrogenesis. In_the future the goal is for more biological replacement of the damaged articular cartilage and stem cells have the potential to deliver this goal.

CONCLUSIONS

The use of stem cells holds promise for the future, but a number of challenges need to be overcome. Further work is needed to identify the optimal source of stem cells for cartilage repair, and also determine conditions that enhance expansion and chondrogenesis. Unfortunately most cell surface markers are inadequate in identifying stem cells unambiguously either because these markers are also expressed by non-stem cells, or they are only expressed by stem cells at a particular stage and under particular culture conditions. The exact nature and location of stem cells in any tissue is also not known. It has recently been suggested that bone marrow derived mesenchymal stem cells originate from microvascular pericytes, and, indeed, many of the tissues from which stem cells have been isolated have good vascularisation and they may give a varied source of cells for future treatments.

Once these challenges have been overcome and these questions answered, we can realistically look at using stem cells in patients and expect consistent and good results.

REFERENCES

[1] Adesida AB, Brady LM, Khan WS, Hardingham TE (2006). The matrix forming phenotype of human meniscus cells is enhanced by expansion in the presence of fibroblast growth factor 2 and hypoxia. *Arthritis Res. Ther.* 8: R61.

[2] Ailhaud G (1982). Adipose cell differentiation in culture. *Mol Cell Biochem.* 49: 17-31.

[3] Akiyama H, Chaboissier MC, Martin JF, Schedl A, de Crombrugghe B (2002). The transcription factor SOX9 has essential roles in successive steps of the chondrocyte differentiation pathway and is required for the expression of SOX5 and SOX6. *Genes Dev.* 16: 2813-28.

[4] Alison MR, Poulsom R, Forbes S, Wright NA (2002). An introduction to stem cells. *J. Path.* 197: 419-23.

[5] Andreeva ER, Pugach IM, Gordon D, Orekhov AN (1998). Continuous subendothelial network formed by pericyte-like cells in human vascular bed. *Tiss Cell.* 30: 127-35.

[6] Baddoo M, Hill K, Wilkinson R, Gaupp D, Hughes C, Kopen GC, Phinney DG (2003). Characterisation of mesenchymal stem cells isolated from murine bone marrow by negative selection. *J. Cell Biochem.* 89: 1235-49.

[7] Bankfalvi A, Terpe HJ, Breukelmann D, Bier B, Rempe D, Schadka G, Krech R, Bocker W (1998). Gains and losses of CD44 expression during breast carcinogenesis and tumour progression. *Histopathology.* 33: 107-16.

[8] Barry FP, Boynton RE, Haynesworth S, Murphy JM, Zaia J (1999). The monoclonal antibody SH-2, raised against human mesenchymal stem cells, recognises an epitope on endoglin (CD105). *Biochem. Biophys. Res. Commun.* 265: 134-9.

[9] Barry F, Boyton R, Murphy M, Haynesworth S, Zaia J (2001). The SH-3 and SH-4 antibodies recognize distinct epitopes on CD73 from human mesenchymal stem cells. Biochem Biophys Res Commun. 289: 519–24.

[10] Bentley G, Minas T (2000). Treating joint damage in young people. *BMJ.* 320: 1585-8.

[11] Bentley G, Biant LC, Carrington RW, Akmal M, Goldberg A, Williams AM, Skinner JA, Pringle J (2003). A prospective, randomised comparison of autologous chondrocyte implantation versus mosaicoplasty for osteochondral defects in the knee. *J. Bone Joint Surg.* 85B: 223-30.

[12] Benya PD, Shaffer JD (1982). Dedifferentiated chondrocytes reexpress the differentiated collagen phenotype when cultured in agarose gels. *Cell;* 30: 215-24.

[13] Beresford JN, Joyner CJ, Devlin C, Triffitt JT (1994). The effects of dexamethasone and 1,25-dihydroxyvitamin D3 on osteogenic differentiation of human marrow stromal cells in vitro. *Arch. Oral. Biol.* 39: 941-7.

[14] Bertram H, Mayer H, Schliephake H (2005). Effect of donor characteristics, technique of harvesting and in vitro processing on culturing of human marrow stroma cells for tissue engineered growth of bone. *Clin. Oral Implants Res.* 16: 524-31.

[15] Bianchi G, Banfi A, Mastrgiacoma M, Notaro R, Luzzatto L, Cancedda R, Quarto R (2003). Ex vivo enrichment of mesenchymal cell progenitors by fibroblast growth factor 2. *Exp. Cell Res.* 287: 98-105.

[16] Bianco P, Riminucci M, Gronthos S, Robey PG (2001). Bone marrow stromal stem cell: nature, biology and potential applications. *Stem Cells.* 19: 180-92.

[17] Brighton CT, Lorich DG, Kupcha R, Reilly TM, Jones AR, Woodbury RA 2nd (1992). The pericyte as a possible osteoblast progenitor cell. *Clin. Orthop. Relat. Res.* 275: 287-99.

[18] Brittberg M, Lindahl A, Nilsson C, Isaksson O, Patterson L (1994). Treatment of deep cartilage defects in the knee with autologous chondrocyte transplantation. *N. Engl. J. Med.* 331: 889-95.

[19] Brittberg M, Tallheden T, Sjogren-Jansson B, Lindahl A, Peterson L (2001). Autologous chondrocytes used for articular cartilage repair: an update. *Clin. Orthop.* 391: S337-48.

[20] Brittberg M, Peterson L, Sjogren-Jansson E, Tallheden T, Lindahl A (2003). Articular cartilage engineering with autologous chondrocyte transplantation. *J. Bone Joint Surg.* 85A: 109-15.

[21] Bruder SP, Jaiswal N, Haynesworth SE (1997a). Growth kinetics, self renewal and the osteogenic potential of purified human mesenchymal stem cells during extensive subcultivation and following cryopreservation. *J. Cell Biochem.* 64: 278-94.

[22] Bruder SP, Horowitz MC, Mosca JD, Haynesworth SE (1997b). Monoclonal antibodies reactive with human osteogenic cell surface antigens. *Bone.* 21: 225-35.

[23] Bruder SP, Ricalton NS, Boynton RE, Connolly TJ, Jaiswal N, Zaia J, Barry FP (1998). Mesenchymal stem cell surface antigen SB-10 corresponds to activated leucocyte cell adhesion molecule and is involved in osteogenic differentiation. *J. Bone Miner Res.* 13: 655-63.

[24] Buckwalter JA (2002). Articular cartilage injuries. *Clin. Orthop. Relat. Res.* 402: 21-37.

[25] Cancedda R, Dozin B, Giannoni P, Quarto R (2003). Tissue engineering and cell therapy of cartilage and bone. *Matrix Biol.* 22: 81-91.

[26] Chia SL, Sawaji Y, Burleigh A, McLean C, Inglis J, Saklatvala J, Vincent T (2009). Fibroblast growth factor 2 is an intrinsic chondroprotective agent that suppresses ADAMTS-5 and delays cartilage degradation in murine osteoarthritis. *Arthritis Rheum.* 60: 2019-27.

[27] Clar C, Cummins E, McIntyre L, Thomas S, Lamb J, Bain L, Jobanputra P, Waugh N (2005). Clinical and cost-effectiveness of autologous chondrocyte implantation for cartilage defects in knee joints: systematic review and economic evaluation. *Health Technol. Assess.* 9: 1-82.

[28] D'Ippolito G, Schiller PC, Ricordi C, Roos BA, Howard GA (1999). Age related osteogenic potential of mesenchymal stromal stem cells from human vertebral bone marrow. *J. Bone Miner Res.* 14: 1115-22.

[29] De Bari C, Dell'Accio F, Tylzanowski P, Luyten FP (2001). Multipotent mesenchymal stem cells from adult human synovial membrane. *Arthritis Rheum.* 44: 1928-42.

[30] Deans RJ, Moseley AB (2000). Mesenchymal stem cells: biology and potential clinical uses. *Exp. Haematol.* 28: 875-84.

[31] Dennis JE, Caplan AI (2000). Bone marrow mesenchymal stem cells. Stem cell handbook by Stewart Sell. Humana Press.

[32] Dennis JE, Carbillet JP, Caplan AI, Charbord P (2002). The STRO-1+ marrow cell population is multipotential. *Cells Tissues Organs.* 170: 73-82.

[33] Deren JA, Kaplan FS, Brighton CT (1990). Alkaline phosphatase production by periosteal cells at various oxygen tensions in vitro. *Clin. Orthop. Rel. Res.* 252: 307-12.
[34] De Wert G, Mummery C (2003). Human embryonic stem cells: research, ethics and policy. *Hum. Reprod.* 18: 672-82.
[35] Devine SM, Hoffman R (2002). Role of mesenchymal stem cells in hematopoietic stem cell transplantation. *Curr. Opin. Hematol.* 7: 358-63.
[36] Doherty M, Ashton B, Walsh S, Beresford J, Grant M, Canfield A (1998). Vascular pericytes express osteogenic potential in vitro and in vivo. *J. Bone Miner Res.* 13: 828-38.
[37] Domm C, Schunke M, Christesen K, Kurz B (2002). Redifferentiation of dedifferentiated bovine articular chondrocytes in alginate culture under low oxygen tension. *Osteoarthritis Cartilage.* 10: 13-22.
[38] Dozin B, Malpeli M, Camardella L, Cancedda R, Pietrangelo A (2002). Response of young, aged and osteoarthritic human articular chondrocytes to inflammatory cytokines: molecular and cellular aspects. *Matrix Biol.* 21: 449-59.
[39] Dragoo JL, Samimi B, Zhu M, Hame SL, Thomas BJ, Lieberman JR, Hedrick MH, Benhaim P (2003). Tissue-engineered cartilage and bone using stem cells from human infrapatellar fat pads. *J. Bone Joint Surg.* 85B: 740-7.
[40] Draper JS, Moore HD, Ruban LN, Gokhale PJ, Andrews PW (2004). Culture and characterization of human embryonic stem cells. *Stem Cells Dev.* 13: 325-36.
[41] Duri ZA, Aichroth PM, Dowd G (1996). The fat pad: Clinical observations. *Am. J. Knee Surg.* 9: 55-66.
[42] Erices A, Conget P, Minguell JJ (2000). Mesenchymal progenitor cells in human umbilical cord blood. *Br. J. Haematol.* 109: 235-42.
[43] Eyre DR (1991). The collagens of articular cartilage. *Semin. Arthritis Rheum.* 21: 2-11.
[44] Farrington-Rock C, Crofts NJ, Doherty MJ, Ashton BA, Griffin-Jones C, Canfield AE (2004). Chondrogenic and adipogenic potential of microvascular pericytes. *Circulation.* 110: 2226-32.
[45] Freidenstein, AJ, Piatetzky II S, Petrakova KV (1966). Osteogenesis in transplants of bone marrow cells. *J. Embryol. Exp. Morphol.* 16: 381-90.
[46] Freidenstein, AJ, Petrakova KV, Kurolesova AI, Frolova GP (1968). Heterotrophic of bone marrow. Analysis of precursoe cells for osteogenic and haematopoietic tissues. *Transplantation.* 6: 230-47.

[47] Freidenstein AJ, Chailakhjan RK, Lalykina KS (1970). The development of fibroblast colonies in monolayer cultures of guinea pig bone marrow and spleen cells. *Cell Tissue Kinet.* 4: 393-403.
[48] Furumatsu T, Tsuda M, Taniguchi N, Tajima Y, Asahara H (2005). Smad3 induces chondrogenesis through the activation of SOX9 via CREB-binding protein/p300 recruitment. *J. Biol. Chem.* 280: 8343-50.
[49] Garcia-Pacheco JM, Oliver C, Kimatrai M, Blanco FJ, Olivares EG (2001). Human decidual stromal cells express CD34 and STRO-1 and are related to bone marrow stromal precursors. *Mol. Hum. Reprod.* 7: 1151-7.
[50] Gregoire FM (2001). Adipocyte differentiation: from fibroblast to endocrine cell. *Exp. Biol. Med.* 226: 997-1002.
[51] Gronthos S, Graves SE, Ohta S, Simmons PJ (1994). The STRO-1+ fraction of adult human boné marrow contains osteogenic precursors. *Blood.* 84: 4164-73.
[52] Gronthos S, Zannettino AC, Graves SE, Ohta S, Hay SJ, Simmons PJ (1999). Differential cell surface expression of the STRO-1 and alkaline phosphatase antigens on discrete developmental stages in primary cultures of human bone cells. *J. Bone Miner Res.* 14: 47-56.
[53] Gronthos S, Zanettino ACW, Shelley JH, Shi S, Graves SE, Kortesidis A, Simmons PJ (2003). Molecular and cellular characterization of highly purified stromal stem cells derived from human bone marrow. *J. Cell Sci.* 116: 1827-35.
[54] Hardingham TE, Fosang AJ (1992). Proteoglycans: many forms and many functions. *FASEB J.* 6: 861-70.
[55] Hardingham TE (1998). Pathophysiology of musculoskeletal disease. In Oxford Textbook of Rheumatology. Maddison PJ, Isenberg DA, Woo P, Glass DN Eds. Second edition Oxford Medical Publications: 1-10.
[56] Hardingham T, Tew S, Murdoch A (2002). Tissue engineering: chondrocytes and cartilage. *Arthritis Res.* 4: S63-8.
[57] Haynesworth SE, Baber MA, Caplan AI (1992). Cell surface antigen on human marrow derived mesenchymal cells are detected by monoclonal antibodies. *Bone.* 13: 69-80.
[58] Herman IM, D'Amore PA (1985). Microvascular pericytes contain muscle and nonmuscle actins. *J. Cell Biol.* 101: 43-52.
[59] Hiraki Y, Shukunami C, Iyama K, Mizuta H (2001). Differentiation of chondrogenic precursor cells during the regeneration of articular cartilage. *Osteoarthritis Cartilage.* 9: S102-8.

[60] Hirschi KK, D'Amore PA (1996). Pericytes in the microvasculature. *Cardiovasc Res.* 32: 687-98.
[61] Hirschi KK, D'Amore PA (1997). Control of angiogenesis by the pericyte: molecular mechanisms and significance. *EXS.* 79: 419-28.
[62] Homicz MR, Schumacher BL, Sah RL, Watson D (2002). Effects of serial expansion of septal chondrocytes on tissue-engineered neocartilage composition. *Otolaryngol Head Neck Surg.* 127: 398-408.
[63] Huang JI, Kazmi N, Durbhakula MM, Hering TM, Yoo JU, Johnstone B (2005). Chondrogenic potential of progenitor cells derived from human bone marrow and adipose tissue: A patient matched comparison. *J. Orthop. Res.* 23: 1383-9.
[64] Hunziker EB (2001). Articular cartilage repair: basic science and clinical progress; a review of the current status and prospects. *Osteoarthritis Cartilage.* 10: 432-63.
[65] Indrawattana N, Chen G, Tadokoro M, Shann LH, Ohgushi H, Tateishi T, Tanaka J, Bunyaratvej A (2004). Growth factor combination for chondrogenic induction from human mesenchymal stem cell. *Biochem. Biophys. Res. Commun.* 320: 914-9.
[66] Ishino T, Hirakawa K, Takeno S, Furukido K, Sugimoto I, Yajin K (2004). Bone-constructing cells from ethmoid bone may have multilineage differentiation potential: preliminary report. *Acta. Otolaryngol Suppl.* 553: 105-8.
[67] Jiang Y, Mishima H, Sakai S, Liu YK, Ohyabu Y, Uemura T (2008). Gene expression analysis of major lineage-defining factors in human bone marrow cells: effect of aging, gender, and age-related disorders. *J. Orthop. Res.* 26: 910-7.
[68] Johnstone B, Hering TM, Caplan AI, Goldberg VM, Yoo JU (1998). In vitro chondrogenesis of bone marrow-derived mesenchymal progenitor cells. *Exp. Cell Res.* 238: 265-72.
[69] Jones EA, Kinsey SE, English A, Jones RA, Straszynski L, Meredith DM, Markham AF, Jack A, Emery P, McGonagle D (2002). Isolation and characterisation of bone marrow multipotential mesenchymal progenitor cells. *Arthritis Rheum.* 46: 3349-60.
[70] Jones EA, English A, Henshaw K, Kinsey SE, Markham AF, Emery P, McGonagle D (2004). Enumeration and phenotypic characterisation of synovial fluid multipotential mesenchymal progenitor cells in inflammatory and degenerative arthritis. *Arthritis Rheum.* 50: 817-27.
[71] Karsenty G, Wagner EF (2002). Reaching a genetic and molecular understanding of skeletal development. *Dev. Cell.* 2: 389-406.

[72] Kato Y, Gospodarowicz D (1985). Sulfated proteoglycan synthesis by confluent cultures of rabbit costal chondrocytes grown in the presence of fibroblast growth factor. *J. Cell Biol.* 100: 477-85.

[73] Khan WS, Adesida AB, Hardingham TE (2007). Hypoxic conditions increase hypoxia-inducible transcription factor 2alpha and enhance chondrogenesis in stem cells from the infrapatellar fat pad of osteoarthritis patients. *Arthritis Res. Ther.* 9: R55.

[74] Khan WS, Tew SR, Adesida AB, Hardingham TE (2008). Human infrapatellar fat pad-derived stem cells express the pericyte marker 3G5 and show enhanced chondrogenesis after expansion in fibroblast growth factor-2. *Arthritis Res. Ther.* 10: R74.

[75] Khan WS, Adesida AB, Tew SR, Hardingham TE (2009). The epitope characterisation and osteogenic differentiation potential of fat pad derived stem cells is maintained with ageing in later life. *Injury.* 40: 150-7.

[76] Khan WS, Adesida AB, Tew SR, Lowe ET, Hardingham TE. Bone Marrow Derived Mesenchymal Stem Cells Express Pericyte Markers in Culture and Show Enhanced Chondrogenesis in Hypoxic Conditions. *J. Orthop.Res.* (in press).

[77] Kim SJ, Min BH, Kim HK (1996). Arthroscopic anatomy of the infrapatellar plica. *Arthroscopy.* 12: 561–4.

[78] Kohn D, Deiler S, Rudert M (1995). Arterial blood supply of the infrapatellar fat pad. Anatomy and clinical consequences. *Arch. Orthop. Trauma Surg.* 114: 72–5.

[79] Koller MR, Bender JG, Papoutsakis ET, Miller WM (1992). Effects of synergistic cytokine combinations, low oxygen, and irradiated stroma on the expansion of human cord blood progenitors. *Blood.* 80: 403-11.

[80] Krampera M, Glennie S, Dyson J, Scott D, Laylor R, Simpson E, Dazzi F (2003). Bone marrow mesenchymal stem cells inhibit the response of naive and memory antigen-specific T cells to their cognate peptide. *Blood.* 101: 3722-9.

[81] Kuroda R, Isada K, Matsumoto T, Akisue T, Fujioka H, Mizuno K, Ohgushi H, Wakitani S, Kurosaka M (2007). Treatment of a full-thickness articular cartilage defect in the femoral condyle of an athlete with autologous bone-marrow stromal cells. *Osteoarthritis Cartilage.* 15: 226-31.

[82] La Prade RF (1998). The anatomy of the deep infrapatellar bursa of the knee. *Am. J. Sports Med.* 26: 129–32.

[83] Le Blanc K, Rasmusson I, Sundberg B, Götherström C, Hassan M, Uzunel M, Ringdén O (2004). Treatment of severe acute graft-versus-host disease with third party haploidentical mesenchymal stem cells. *The Lancet.* 363: 1439-41.
[84] Lee CR, Grodzinsky AJ, Hsu HP, Martin SD, Spector M (2000). Effects of harvest and selected cartilage repair procedures on the physical and biochemical properties of articular cartilage in the canine knee. *J. Orthop. Res.* 18: 790-9.
[85] Lennon DP, Edmison JM, Caplan AI (2001). Cultivation of rat marrow-derived mesenchymal stem cells in reduced oxygen tension: effects on in vitro and in vivo osteochondrogenesis. *J. Cell Physiol.* 187: 345-55.
[86] Li Y, Tew SR, Russel AM, Gonzalez KR, Hardingham TE, Hawkins RE (2004). Transduction of passaged human articular chondrocytes with adenoviral, retroviral, and lentiviral vectors and the effects of enhanced expression of SOX9. *Tissue Eng.* 10: 575-84.
[87] Loty S, Forest N, boulekbache H, Sautier JM (1995). Cytochalasin D induces changes in cell shape and promotes in vitro chondrogenesis: a morphological study. *Biol. Cell.* 83: 149-61.
[88] MacDougald OA, Mandrup S (2002). Adipogenesis: forces that tip the scales. Trends Endocrinol Metab. 13: 5-11.
[89] Mackay AM, Beck SC, Murphy JM, Barry FP, Chichester CO, Pittenger MF (1998). Chondrogenic differentiation of cultured marrow mesenchymal stem cells from marrow. *Tissue Eng.* 4: 415-28.
[90] Magne D, Vinatier C, Julien M, Weiss P, Guicheux J (2005a). Mesenchymal stem cell therapy to rebuild cartilage. *Trends Mol. Med.* 11: 519-26.
[91] Magne D, Julien M, Vinatier C, Merhi-Soussi F, Weiss P, Guicheux J (2005b). Cartilage formation in growth plate and arteries: from physiology to pathology. *Bioessays;* 27: 708-16.
[92] Majumdar MK, Keane-Moore M, Buyaner D, Hardy WB, Moorman MA, McIntosh KR, Mosca JD (2003). Characterization and functionality of cell surface molecules on human mesenchymal stem cells. *J. Biomed. Sci.* 10: 228-41.
[93] Manne U, Srivastava RG, Srivastava S (2005). Recent advances in biomarkers for cancer diagnosis and treatment. *Drug Discov. Today.* 10: 965-76.
[94] Mareschi K, Ferrero I, Rustichelli D, Aschero S, Gammaitoni L, Aglietta M, Madon E, Fagioli F (2006). Expansion of mesenchymal stem cells

isolated from paediatric and adult donor bone marrow. *J. Cell Biochem.* 97: 744-54.

[95] Martin I, Muraglia A, Campanile G, Cancedda R, Quarto R (1997). Fibroblast growth factor-2 supports ex vivo expansion and maintenance of osteogenic precursors from human bone marrow. *Endocrinology.* 138: 4456-62.

[96] Martin JA, Buckwalter JA (2003). The role of chondrocyte senescence in the pathogenesis of osteoarthritis and in limiting cartilage repair. *J. Bone Joint Surg.* 85A: S106-10.

[97] Meyrick B, Fujiwara K, Reid L (1981). Smooth muscle myosin in precursor and mature smooth muscle cells in normal pulmonary arteries and the effect of hypoxia. *Exp. Lung Res.* 2: 303-13.

[98] Miranville A, Heeschen C, Sengenes C, Curat CA, Busse R, Bouloumie A (2004). Improvement of postnatal neovascularisation by human adipose tissue derived stem cells. *Circulation.* 110: 349-55.

[99] Mochizuki T, Muneta T, Sakaguchi Y, Nimura A, Yokoyama A, Koga H, Sekiya I (2006). Higher chondrogenic potential of fibrous synovium- and adipose synovium- derived cells compared with subcutaneous fat-derived cells. *Arthritis Rheum.* 54: 843-53.

[100] Morrison SJ, Csete M, Groves AK, Melega W, Wold B, Anderson DJ (2000). Culture in reduced levels of oxygen promotes clonogenic sympathoadrenal differentiation by isolated neural crest stem cells. *J. Neurosci.* 20: 7370-6.

[101] Murphy CL, Sambanis A (2001). Effect of oxygen tension and alginate encapsulation on restoration of the differentiated phenotype of passaged chondrocytes. *Tissue Eng.* 7: 791-803.

[102] Nayak RC, Berman AB, George KL, Eisenbarth GS, King GL (1988). A monoclonal antibody (3G5) defined ganglioside antigen is expressed on the cell surface of microvascular pericytes. *J. Exp. Med.* 167: 1003-15.

[103] Newman A (1998). Articular cartilage repair. *Am. J. Sports Med.* 26: 309-24.

[104] O'Driscoll SW (1998). The healing and regeneration of articular cartilage. *J. Bone Joint Surg. Am.* 80: 1795-812.

[105] Ogilvie-Harris DJ, Giddens J (1994). Hoffa's disease: arthroscxopic resection of the infrapatellar fat pad. *Arthroscopy.* 10: 184-7.

[106] Orkin SH, Morrison SJ (2002). Stem-cell competition. *Nature.* 418:25-7.

[107] Owen ME, Cave J, Joyner CJ (1987). Clonal analysis in vitro of osteogenic differentiation of marrow CFU-F. *J. Cell Sci.* 87: 731-8.

[108] Owen M, Friedenstein AJ (1988). Stromal stem cells: marrow-derived osteogenic precursors. *Ciba Found Symp.* 136: 42-60.
[109] Parsons IM, Sonnabend DH (2004). What is the role of joint replacement surgery? *Best Pract Res. Clin. Rheumatol.* 18: 557-72.
[110] Peng H, Huard J (2003). Stem cells in the treatment of muscle and connective tissue diseases. *Curr. Opin. Pharmacol.* 3: 329-33.
[111] Peterson L, Minas T, Brittberg M, Lindahl A (2003). Treatment of osteochondritis dissecans of the knee with autologous chondrocyte transplantation: results at two to ten years. *J. Bone Joint Surg.* 85A: 17-24.
[112] Pittenger MF, Mackay AM, Beck SC, Jaiswal RK, Douglas R, Mosca JD, Moorman MA, Simonetti DW, Craig S, Marshak DR (1999). Multilineage potential of adult human mesenchymal stem cells. *Science.* 284: 143-7.
[113] Pittenger MF, Mbalaviele G, Black M, Mosca JD, Marshak DR (2001). Mesenchymal stem cells. In: Koller MR, Palsson BO, Masters JRW Eds. Primary mesenchymal cells. Dordrecht; Boston: Kluwer Academic Publishers: 189-207.
[114] Pittinger MF, Martin BJ (2004). Mesenchymal stem cells and their potential as cardiac therapeutics. *Circ. Res.* 95: 9-20.
[115] Poole AR (1995). Imbalance of anabolism and catabolism of cartilage matrix componenets in osteoarthritis. In: Kuettner KE, Goldberg B Eds. Osteoarthritic disorders. Rosemont: AAOS: 247-60.
[116] Prockop DJ (1997). Marrow stromal cells as stem cells for nonhaematopoietic tissues. *Science.* 276: 71-4.
[117] Qiu Z, Wei Y, Chen N, Jiang M, Wu J, Liao K (2001). DNA synthesis and mitotic clonal expansion is not a required step for 3T3–L1 preadipocyte differentiation into adipocytes. *J. Biol. Chem.* 276: 11988-95.
[118] Quirici N, Soligo D, Bossolasco P, Servida F, Lumini C, Deliliers GL (2002). Isolation of bone marrow mesenchymal stem cells by anti-nerve growth factor receptor antibodies. *Exp. Haematol.* 30: 783-91.
[119] Raghunath J, Salacinski HJ, Sales KM, Butler PE, Seifalian M (2005). Advancing cartilage tissue engineering: the application of stem cell technology. *Curr. Opin. Biotech.* 16: 503-9.
[120] Ramirez-Zacarias JL, Castro-Munozledo F, Kuri-Harcuch W (1992). Quantitation of adipose conversion and triglycerides by staining intracytoplasmic lipids with Oil red O. *Histochem.* 97: 493-7.

[121] Rhodin JA (1968). Ultrastructure of mammalian venous capillaries, venules, and small collecting veins. *J. Ultrastruct. Res.* 25: 452-500.
[122] Rich IN, Kubanek B (1982). The effect of reduced oxygen tension on colony formation of erythropoietic cells in vitro. *Br. J. Haematol.* 52: 579-88.
[123] Rich IN (1986). A role for the macrophage in normal hemopoiesis. II. Effect of varying physiological oxygen tensions on the release of hemopoietic growth factors from bone marrow derived macrophages in vitro. *Exp. Haematol.* 14: 746-51.
[124] Rim JS, Mynatt RL, Gawronska-Kozak B (2005). Mesenchymal stem cells from the outer ear: a novel adult stem cell model system for the study of adipogenesis. *FASEB J.* 19: 1205-7.
[125] Risbud MV, Sittinger M (2002). Tissue engineering: advances in in vitro cartilage generation. *Trends Biotech.* 20: 351-6.
[126] Rosen ED, Spiegelman BM (2000). Molecular regulation of adipogenesis. *Annu. Rev. Cell Dev. Biol.* 16: 145-71.
[127] Roura S, Farré J, Soler-Botija C, Llach A, Hove-Madsen L, Cairó JJ, Gòdia F, Cinca J, Bayes-Genis A (2006). Effect of aging on the pluripotential capacity of human CD105+ mesenchymal stem cells. *Eur. J. Heart Fail.* 8: 555-63.
[128] Saddik D, McNally EG, Richardson M (2004). MRI of Hoffa's fat pad. *Skeletal Radiol.* 33: 433-44.
[129] Sakaguchi Y, Sekiya I, Yagishita K, Muneta T (2005). Comparison of human stem cells derived from various mesenchymal tissues- Superiority of synovium as a cell source. *Arthritis Rheum.* 52: 2521-9.
[130] Scharstuhl A, Schewe B, Benz K, Gaissmaier C, Bühring HJ, Stoop R (2007). Chondrogenic potential of human adult mesenchymal stem cells is independent of age or osteoarthritis etiology. *Stem Cells.* 25: 3244-51.
[131] Scherer K, Schunke M, Sellckau R, Hassenpflug J, Kurz B (2004). The influence of oxygen and hydrostatic pressure on articular chondrocytes and adherent bone marrow cells in vitro. *Biorheology.* 41: 323-33.
[132] Schipani E, Ryan HE, Didrickson S, Kobayashi T, Knight M, Johnson RS (2001). Hypoxia in cartilage: HIF1 is essential for chondrocyte growth arrest and survival. *Genes Dev.* 15: 2865-76.
[133] Schipper BM, Marra KG, Zhang W, Donnenberg AD, Rubin JP (2008). Regional anatomic and age effects on cell function of human adipose-derived stem cells. *Ann. Plast. Surg.* 60: 538-44.

[134] Schulz RM, Bader A (2007). Cartilage tissue engineering and bioreactor systems for the cultivation and stimulation of chondrocytes. *Eur. Biophys. J.* 36: 539-68.
[135] Sekiya I, Vuoristo JT, Larson BL, Prockop DJ (2002). In vitro cartilage formation by human adult stem cells from bone marrow stroma defines the sequence of cellular and molecular events during chondrogenesis. *Proc. Natl. Acad. Sci. USA.* 99: 4397-402.
[136] Shamsul BS, Aminuddin BS, Ng MH, Ruszymah BH (2004). Age and gender effect on the growth of bone marrow stromal cells in vitro. *Med. J. Malaysia.* 59B:196-7.
[137] Shapiro F, Koide S, Glimcher MJ (1993). Cell origin and differentiation in the repair of full thickness defects of articular cartilage. *J. Bone Joint Surg. Am.* 75: 532-53.
[138] Shi S, Gronthos S (2003). Perivascular niche of postnatal mesenchymal stem cells in human bone marrow and dental pulp. *J. Bone Miner Res.* 18: 696-704.
[139] Short B, Brouard N, Occhioduro-Scott T, Ramakrishnand A, Simmons PJ (2003). Mesenchymal stem cells. *Arch. Med. Res.* 34: 565-71.
[140] Siddappa R, Licht R, van Blitterswijk C, de Boer J (2007). Donor variation and loss of multipotency during in vitro expansion of human mesenchymal stem cells for bone tissue engineering. *J. Orthop. Res.* 25: 1029-41.
[141] Silver IA (1975). Measurement of pH and ionic composition of pericellular sites. *Philos Trans. R Soc. Lond B Biol. Sci.* 271: 261-72.
[142] Simmons PJ, Torok-Storb B (1991). Identification of stromal cell prcursors in human bone marrow by a novel monoclonal antibody, STRO-1. *Blood.* 78: 55-62.
[143] Simon L (1999). Osteoarthritis: a review. *Clin. Cornerstone.* 2: 26-37.
[144] Smillie IS (1974). Disease of the knee joint, 1st edn. London: Churchill Livingstone.
[145] Solchaga LA, Penick K, Porter JD, Goldberg VM, Caplan AI, Welter JF (2005). FGF-2 enhances the mitotic and chondrogenic potentials of human adult bone marrow-derived mesenchymal stem cells. *J. Cell Physiol.* 203: 398-409.
[146] Sottile V, Seuwen K (2001). A high capacity screen for adipogenic differentiation. *Anal. Biochem.* 293: 124-8.
[147] Spees JL, Gregory CA, Singh H, Tucker HA, Peister A, Lynch PJ, Hsu SC, Smith J, Prockop DJ (2004). Internalised antigens must be removed

to prepare hypoimmunogenic mesenchymal stem cells for cell and gene therapy. *Mol. Ther.* 9: 747-56.
[148] Stenderup K, Justesen J, Clausen C, Kassem M (2003). Aging is associated with decreased maximal life span and accelerated senescence of bone marrow stromal cells. *Bone.* 33: 919-26.
[149] Stewart K, Walsh S, Screen J, Jefferiss CM, Chainey J, Jordan GR Beresford JN (1999). Further characterisation of cells expressing STRO-1 in cultures of adult human bone marrow stromal cells. *J. Bone Miner Res.* 14: 1345-56.
[150] Stolzing A, Jones E, McGonagle D, Scutt A (2008). Age-related changes in human bone marrow-derived mesenchymal stem cells: consequences for cell therapies. *Mech. Ageing Dev.* 129: 163-73.
[151] Suva D, Garavaglia G, Menetrey J, Chapuis B, Hoffmeyer P, Bernheim L, Kindler V (2004). Non-hematopoietic human bone marrow contains long-lasting, pluripotential mesenchymal stem cells. *J. Cell Physiol.* 198: 110-8.
[152] Tew SR, Hardingham TE (2006). Regulation of SOX9 mRNA in Human Articular Chondrocytes Involving p38 MAPK Activation and mRNA Stabilization. *J. Biol. Chem.* 281: 39471-9.
[153] Thomas WE (1999). Brain macrophages: on the role of pericytes and perivascular cells. *Brain Res. Brain Res. Rev.* 31: 42-57.
[154] Thomson JA, Itskovitz-Eldor J, Shapiro SS, Waknitz MA, Swiergiel JJ, Marshall VS, Jones JM (1998). Embryonic stem cell lines derived from human blastocysts. *Science.* 282: 1145-7.
[155] Triffitt JT (2002). Stem cells and the philosopher's stone. *J. Cell Biochem. Suppl.* 38: 38: 13-9.
[156] Vaananen HK (2005). Mesenchymal stem cells. Ann Med. 37: 469-79.
[157] Vats A, Tolley NS, Bishop AE, Polak JM (2005). Embryonic stem cells and tissue engineering: delivering stem cells to the clinic. *J. R. Soc. Med.* 98: 346-50.
[158] Wakitani S, Goto T, Pineda SJ, Young RG, Mansour JM, Caplan AI, Goldberg VM (1994). Mesenchymal cell-based repair of large, full-thickness defects of articular cartilage. *J. Bone Joint Surg. Am.* 76: 579-92.
[159] Wang DW, Fermor B, Gimble JM, Awad HA, Guilak F (2005). Influence of oxygen on the proliferation and metabolism of adipose derived adult stem cells. *J. Cell Physiol.* 204: 184-91.

[160] Watt FM (1988). Effect of seeding density on stability of the differentiated phenotype of pig articular chondrocytes in culture. *J. Cell Sci.* 89: 373-8.

[161] Wesseling P, Schlingemann RO, Rietveld FJ, Link M, Burger PC, Ruth DJ (1995). Early and extensive contribution of pericytes/ vascular smooth muscle cells to microvascular proliferation in glioblastoma multiforme: an immuno-light and immuno-electron microscopy study. *J. Neuropathol. Exp. Neurol.* 54: 304-10.

[162] Wickham MQ, Erickson GR, Gimble JM, Vail TP, Guilak F (2003). Multipotent stromal cells derived from the infrapatellar fat pad of the knee. *Clin. Orthop.* 412: 196-212.

[163] Williams P, Warwick R, Dyson M et al (1989). Gray's Anatomy, 37th edn. New York: Churchill Livingstone.

[164] Wollheim F (1996). Current pharmacological treatments of osteoarthritis. *Drugs*. 52: 27-38.

[165] Wroblewski J, Edwall-Arvidsson C (1995). Inhibitory effects of basic fibroblast growth factor on chondrocyte differentiation. *J. Bone Miner Res.* 10: 735-42.

[166] Zhou G, Garofalo S, Mukhopadhyay K, Lefebvre V, Smith CN, Eberspaecher H, de Crombrugghe B (1995). A 182 bp fragment of the mouse pre alpha 1 (II) collagen gene is sufficient to direct chondrocyte expression in transgenic mice. *J. Cell Sci.* 108: 3677-84.

[167] Zuk PA, Zhu M, Mizino H, Huang J, Futrell JW, Katz AJ, Benhaim P, Lorenz HP, Hedrick MH (2001). Multilineage cells from human adipose tissue: implications for cell-based therapies. *Tissue Eng.* 7: 211-28.

[168] Zuk PA, Zhu M, Ashjian P, De Ugarte DA, Huang JI, Mizino H, Alfonso ZC, Fraser JK, Benhaim P, Hedrick MH (2002). Human adipose tissue is a source of multipotent stem cells. *Mol. Biol. Cell.* 13: 4279-95.

[169] Stolzing A, Jones E, McGonagle D, Scutt A (2008). Age-related changes in human bone marrow-derived mesenchymal stem cells: consequences for cell therapies. *Mech. Ageing Dev.* 129: 163-73.

[170] Suva D, Garavaglia G, Menetrey J, Chapuis B, Hoffmeyer P, Bernheim L, Kindler V (2004). Non-hematopoietic human bone marrow contains long-lasting, pluripotential mesenchymal stem cells. *J. Cell Physiol.* 198: 110-8.

INDEX

A

acid, 21
activation, 14, 22, 39
adhesion, 14, 37
adipocyte, 10
adipose, ix, 5, 6, 7, 11, 40, 43, 44, 45, 47, 48
adipose tissue, ix, 5, 6, 7, 11, 40, 43, 48
adult stem cells, ix, 46, 47
adventitia, 17
age, 2, 7, 8, 23, 26, 40, 45
ageing, 7, 16, 41
agent, 37
aggregates, 9
aggregation, 22
aging, 8, 40, 45
aging population, 8
alternative, 7, 29, 30
anabolism, 44
anatomy, 41
angiogenesis, 40
antibody, 13, 14, 18, 19
antigen, 2, 14, 37, 39, 41, 43
apoptosis, 2, 22, 23
arrest, 45
arteries, 6, 17, 18, 42
arterioles, 18
arthritis, 23
arthrodesis, 25
arthroplasty, 7, 25
arthroscopy, 7
arthrotomy, 26
articular cartilage, x, 22, 23, 24, 26, 29, 30, 31, 36, 38, 39, 41, 42, 43, 46, 47
aspirate, 6, 13, 15
availability, 25

B

basal lamina, 18
basement membrane, 17, 18
binding, 22, 39
biomarkers, 42
biopsy, 7, 26
biosynthesis, 10
blood, 5, 6, 11, 18, 38, 41
blood flow, 18
blood supply, 6, 41
bone cells, 39
bone marrow, vii, ix, 2, 3, 5, 6, 7, 11, 12, 13, 14, 15, 16, 17, 18, 19, 23, 24, 25, 29, 30, 33, 35, 37, 38, 39, 40, 43, 44, 45, 46, 47, 48
bursa, 41

C

calcification, 18

cancer, 42
capillary, 17
carbon, 11
carbon dioxide, 11
carcinogenesis, 35
cartilage, vii, ix, 7, 9, 11, 12, 21, 22, 23, 24, 25, 26, 29, 30, 33, 36, 37, 38, 39, 40, 42, 43, 44, 45, 46
cartilaginous, 22
catabolism, 44
cell cycle, 10
cell surface, ix, 6, 7, 8, 13, 14, 16, 19, 33, 37, 39, 42, 43
chondrocyte, x, 7, 9, 22, 27, 29, 35, 36, 37, 43, 44, 45, 48
chronic obstructive pulmonary disease, 3
clinical trials, 3
coding, 31
collagen, 9, 11, 21, 22, 23, 25, 26, 30, 36, 48
competition, 43
complications, 8, 25
components, 11, 17
composition, 40, 46
condensation, 9, 23
connective tissue, 6, 21, 44
contamination, 6
control, 11, 18, 23, 30
control group, 30
conversion, 44
COPD, 3
costs, 23
cryopreservation, 37
cultivation, 46
culture conditions, x, 2, 8, 9, 13, 16, 30, 33
cycles, 10
cycling, 1
cytokines, 24, 38

D

daily living, 23
debridement, 26
defects, 7, 23, 24, 26, 29, 30, 36, 37, 46, 47
degenerative arthritis, 40

degradation, 37
delivery, 30, 31
density, 9, 11, 22, 48
developed countries, 23
developmental process, 22
differentiation, ix, x, 1, 3, 5, 7, 8, 9, 10, 12, 18, 22, 23, 29, 30, 35, 36, 37, 39, 40, 41, 42, 43, 44, 46, 48
disease progression, 24
disorder, 23
distribution, 18
division, 1

E

earnings, 23
ectoderm, 1
elderly, 25
electron microscopy, 48
embryo, 1
embryonic stem cells, ix, 1, 2, 3, 38
encapsulation, 43
endocrine, 39
endoderm, 1
endothelial cells, 14, 16, 17, 18
endothelium, 17
environment, 9
enzymes, 11
epithelial cells, 14, 17
ethics, 38
etiology, 45
examinations, 8
exercise, 24
extracellular matrix, 11, 21, 23

F

failure, 23, 24
family, 23
fat, ix, 5, 6, 7, 8, 11, 12, 13, 15, 16, 17, 30, 38, 41, 43, 45, 48
fibrin, 26
fibroblast growth factor, 35, 36, 41, 48
fibroblasts, 3, 16

fibrosis, 7
fibrous tissue, 6, 25
fixation, 1

G

gel, 30
gender, 40, 46
gene expression, 9, 26
gene therapy, 47
generation, 45
germ layer, 1
glioblastoma multiforme, 48
glucose, 9
glycerol, 10
glycine, 21
grading, 30
growth, 9, 21, 22, 24, 31, 36, 37, 42, 43, 45, 46
growth factor, 9, 24, 31, 37, 43, 45

H

harvesting, 36
healing, 43
health, 26
hepatocytes, 3
hip, 23, 26
hip replacement, 23
histocompatability, 3
host, 3, 42
hyaline, 21, 25, 27, 30
hypoxia, 11, 35, 41, 43

I

in vitro, 2, 3, 9, 11, 15, 16, 18, 36, 38, 42, 43, 45, 46
in vivo, 9, 11, 12, 18, 38, 42
inducer, 10
induction, 40
inflammation, 24
inflammatory bowel disease, 3
infrapatellar, 5, 6, 38, 41, 43, 48

injuries, 29, 37
insulin, 10, 24
integrity, 18, 21
interactions, 21
intima, 17
ions, 22
isolation, 8, 30

J

joint damage, 36
joints, 22, 24, 25, 26, 37

K

karyotype, 2
kidney, 5
kinetics, 37
knee arthroplasty, 7

L

later life, 8, 16, 41
lesions, 1, 25
leucocyte, 2, 37
life expectancy, 23
life span, 47
lifestyle, 25
lifetime, 25
ligament, 6
likelihood, 25
lipids, 44
liver, 5
lumen, 17
lymphocytes, 14

M

macrophages, 45, 47
management, 23, 24, 30
marrow, 5, 6, 7, 13, 15, 16, 18, 30, 36, 37, 39, 41, 42, 43, 44
matrix, 9, 11, 12, 21, 22, 24, 35, 44

maturation, 9, 23
media, 6, 17, 18
medication, 24
membranes, 1
memory, 41
mesenchymal stem cells, vii, ix, 1, 2, 7, 10, 14, 16, 29, 30, 33, 35, 36, 37, 38, 41, 42, 44, 45, 46, 47, 48
mesenchyme, 3, 18
mesoderm, 1, 3, 11
metabolism, 47
mice, 48
migration, 18
mitogen, 11
model system, 45
models, 2
molecules, 3, 21, 22, 30, 42
monoclonal antibody, 14, 35, 43, 46
monolayer, 10, 16, 26, 39
morbidity, 5, 6, 7, 26
morphology, 11
mRNA, 11, 47
multipotent, 2, 48
musculoskeletal system, 23
myoblasts, 3
myosin, 43

N

National Health Service, 23
necrosis, 24
nerve growth factor, 14, 15, 44
network, 18, 21, 24, 35

O

observations, 38
occupational therapy, 24
osmotic pressure, 22
ossification, 22
osteoarthritis, 23, 24, 25, 37, 41, 43, 44, 45, 48
osteochondritis dissecans, 44
osteotomy, 30

oxygen, 11, 38, 41, 42, 43, 45, 47

P

pain, 6, 24, 25
paradigm shift, 1
patella, 6
pathogenesis, 43
pathology, 42
pathways, 10
periosteum, 5, 7, 26
peripheral blood, 5
permeability, 18
pharmacological treatment, 48
phenotype, 6, 8, 11, 35, 36, 43, 48
physiology, 42
placenta, 1
platelets, 14
population, 1, 3, 10, 18, 19, 23, 37
precursor cells, 16, 39
pressure, 11, 45
production, 1, 24, 38
proliferation, 1, 3, 6, 7, 8, 10, 12, 26, 47, 48
propagation, 1
prostheses, 25
prosthesis, 7
proteins, 21, 24, 26, 31
proteoglycans, 21, 22
pulmonary arteries, 43
pulp, 5, 18, 46

R

reactivity, 16
recombinant DNA, 31
regeneration, 1, 3, 39, 43
regulation, 45
rejection, 3
repair, vii, ix, 2, 3, 7, 8, 12, 16, 23, 24, 25, 26, 29, 30, 33, 36, 40, 42, 43, 46, 47
resection, 7, 43
retina, 5
risk, 1, 5, 8

S

safety, 2
scores, 30
secrete, 21, 22
seeding, 48
senescence, 43, 47
sepsis, 6
serum, 16
shape, 42
side effects, 7
signalling, 22, 23, 30
signals, 22
signs, 23, 29
skeletal muscle, 5, 7
skin, 5
smooth muscle cells, 13, 17, 18, 43, 48
somatic cell, 2
spindle, 2
spleen, 5, 39
stability, 18, 48
stages, 2, 6, 9, 10, 39
stem cell lines, 47
strategies, 1, 7, 24, 26, 29
stroma, 36, 41, 46
stromal cells, 5, 15, 36, 39, 41, 44, 46, 47, 48
surgical intervention, 25
survival, 10, 45
swelling, 24
synergistic effect, 9
synovial fluid, 40
synovial membrane, 37
synovial tissue, 5, 6, 16
synthesis, 11, 41, 44

T

T cell, 41
tension, 11, 38, 42, 43, 45
TGF, 9, 12, 13, 14, 22, 24
therapeutics, 44
therapy, 37, 42
thymus, 5
tissue, vii, ix, x, 1, 2, 3, 5, 6, 7, 8, 9, 11, 17, 18, 21, 22, 24, 26, 29, 30, 33, 36, 40, 44, 46, 47
tissue remodelling, 2
TNF, 24
transcription, 10, 35, 41
transforming growth factor, 22
transmission, 3
transplantation, 25, 36, 38, 44
trauma, 23
trial, 24, 26, 30
triglycerides, 10, 44
trypsin, 14
turnover, 21, 24
tyrosine, 14

V

variability, 10, 16
variation, 46
vasculature, 14
venules, 17, 18, 45
vessels, 6, 17

W

wealth, 5
wear, 21
workers, 2
wound infection, 6